SECURE BY DESIGN

How Modern Organizations Collaborate Without Compromise

The Microsoft 365 playbook for M&A, due diligence, and sensitive external work – without a standalone VDR platform.

First edition.

ISBN
Paperback: 979-8-90190-463-3
Hardcover: 979-8-90190-464-0

Dedication

This book is dedicated to three groups of people without whom neither the work nor the thinking behind it would exist.

First, to my previous and current team at **Netwoven.** Your curiosity, discipline, and willingness to challenge assumptions have shaped the ideas in this book far more than you may realize. Building real solutions in the real world requires rigor, humility, and persistence, and I am grateful to work alongside people who bring those qualities every day.

Second, to my **family**.

Thank you for your patience, encouragement, and quiet support- the kind that makes long hours and deep thinking possible. Your belief in me, even when the work pulls me away, is something I never take for granted.

And finally, to my **friends and mentors**.

You've pushed my thinking, broadened my perspective, and reminded me that good work is rarely done alone. Your conversations, questions, and candor have helped sharpen both the ideas and the intent behind this book.

This work is as much yours as it is mine.

Foreword

By Indranil Sircar

Global CTO, Manufacturing & Mobility Industry, Microsoft

In every modern organization, the work that matters most is also the work that carries the greatest risk. Deals, audits, board reporting, regulatory reviews, and cross-company partnerships all demand rapid collaboration - yet they also demand precision, confidentiality, and a defensible record of what happened.

For years, the default answer was to rely on a Virtual Data Room. It served its purpose when sensitive collaboration was occasional and contained. But as these processes became continuous, the traditional VDR model began to show its limits: duplicated data, siloed workflows, unpredictable costs, and a growing disconnect from the systems organizations already trust.

Secure by Design reframes this challenge with clarity. It argues that the real need was never the VDR itself, but the outcomes it promised: controlled access, secure external collaboration, protection that travels with the file, auditability, time-bound access, clean close-out, and external credibility. And it shows that these outcomes can be delivered inside Microsoft 365, without exporting sensitive work into yet another platform.

What makes this book especially valuable is its practicality. Niraj Tenany brings years of experience working with business leaders, IT teams, security architects, and compliance officers, all of whom face the same tension: enable the business to move fast, without compromising control. His model is simple, repeatable, and defensible - a way to collaborate without compromise.

This is not a book about tools. It is a playbook for leaders who must protect every deal, every discussion, and every decision. In a world where collaboration is unavoidable and compromise is unacceptable; this book offers a clear path forward.

Preface

This book started with a simple observation: when organizations say, *"We need a Virtual Data Room,"* they're rarely asking for a tool.

They're asking for confidence.

Confidence that sensitive information isn't drifting.

Confidence that external collaboration won't create hidden risk. And confidence that, later, they can prove what happened-without reconstruction or guesswork.

Over the years of working with business leaders, IT teams, and security stakeholders within Microsoft 365, I saw the same pattern recur. The challenge wasn't a lack of technology, but the lack of **clear outcomes and a defensible operating model**.

When VDRs are treated as products, decisions collapse into feature comparisons. When they're treated as outcomes, better questions emerge: How do we control access without slowing the business? How does protection stay with the data? And how do we close cleanly-with proof?

This book is about answering those questions in practical terms.

It presents a way to deliver VDR-grade outcomes-secure access, auditability, and confidence-inside Microsoft 365, without creating another silo or exception. The focus is not on tools, but on principles, lifecycle discipline, and repeatability.

If you're responsible for sensitive external collaboration under real pressure, I hope this book gives you clarity-and a model that holds up when it matters most.

- **Niraj Tenany**

Table of Contents

Introduction

Most organizations don't want a "Virtual Data Room." They want the outcomes that a VDR is supposed to deliver - especially when sharing highly sensitive information with external parties.

And when buyers say, "We need a VDR," they're usually not asking for a portal, a vendor, or a feature checklist. They're describing a consistent set of expectations:

- controlled access
- secure collaboration with external parties
- clear visibility into who accessed what and when
- confidentiality by default
- time-bound access with clean closeout
- external credibility.

Those expectations exist regardless of company size or deal type.

This book starts with a simple reframing:

A VDR is not a product. A VDR is an outcome.

That distinction matters because it changes how you evaluate solutions. If you treat a VDR as 'a tool', you end up comparing feature lists. If you treat it as an outcome standard, you focus on what the business truly needs: control, auditability, confidence, and a clean close-out. All delivered reliably, even under pressure.

Why does this book exist?

Sharing sensitive information externally is harder and riskier than it should be. Organizations still rely on email, links, and ad-hoc file sharing, or they swing to expensive standalone VDRs that introduce yet another silo. Meanwhile, business teams want speed, IT wants standardization, and security wants control and auditability-and these needs are often in conflict.

If you already run on Microsoft 365, there's an even sharper tension: many traditional VDR approaches require moving sensitive information into third-party environments. That may have been acceptable years ago. But today, the default Microsoft 365 approach (links + basic sharing) also has well-known failure modes: links shared too broadly, guests lingering long after projects end, and unclear visibility into who still has access.

This book is written for that reality: **external collaboration is constant**, the stakes are high, and 'secure enough for the deal' is not the same as 'defensible as a standard.'

Who this book is for

This book is for leaders and operators who are accountable for high-stakes collaboration outcomes, including:

- Business teams running deals, audits, and sensitive collaborations who need confidence and speed.
- IT and digital workplace leaders who need secure external collaboration to be standardized on Microsoft 365, not reinvented per department.
- Security and compliance teams who need policy enforcement, visibility, and audit readiness, without creating a parallel "deal security" perimeter.
- Legal and finance teams who care about confidentiality and defensibility in high-stakes processes.

It is also for anyone who is trying to answer a practical question: **"How do we deliver VDR-grade outcomes inside Microsoft 365-without turning every deal into a new exception?"**

The one-sentence thesis

You can deliver true VDR outcomes inside Microsoft 365-secure access, auditability, and confidence-without a standalone VDR platform, by applying the right control layer and operating model in place.

What this book will-and won't-do

What it *will* do:

- Define the VDR outcome standard in plain terms and give you a checklist you can use to evaluate any approach.
- Explain why traditional VDR models can solve a narrow problem but create broader governance and repeatability issues (cost, silos, duplicated data, poor M365 integration, and one-off workflows).
- Provide a lifecycle operating model designed to be defensible-not just during the deal, but across setup, collaboration, evidence, and closeout.

What it *won't* do:

- It will not be a product manual for SharePoint, Purview, or any specific vendor. The goal is to help you **reason about outcomes, controls, and repeatability**, then choose or design an implementation that fits your environment.
- It will not pretend that 'turning SharePoint into a VDR' is just a folder exercise. Transforming Microsoft 365 components into a functional VDR requires governance discipline and configuration choices; the book focuses on the operating pattern that makes it work.

How to use this book (practical reading paths)

You can read straight through, but it's designed to be useful in three common ways:

1. **If you're evaluating options right now.**
   Read Part I and Appendix A first. You'll get the outcome standard and the questions that separate "feature lists" from "defensible outcomes."

2. **If you need to run a live deal/audit/fundraise in Microsoft 365.**
   Skim Part II and Appendix B. It gives you the lifecycle pattern you can apply immediately-setup → collaboration → evidence → close-out.
3. **If you're trying to standardize across the enterprise.**
   Read Part III after Part II. It helps you communicate the same core story differently to business buyers vs. IT/security buyers, without changing the underlying outcome standard.

How the book is organized

The structure is intentionally compact and outcome-driven:

Part I - The Reframe

- Chapter 1: VDR Outcomes, Not VDR Tools
- Chapter 2: Why Traditional VDRs Fail at Scale
- Chapter 3: The Microsoft 365–Native Alternative (The Thesis)

Part II - The Defensible Operating Model

- Chapter 4: The 4-Phase Lifecycle That Makes It Defensible
- Chapter 5: Identity & External Access: Collaborate Without Losing Control
- Chapter 6: Protection That Travels with the File
- Chapter 7: Evidence & Auditability: Turn Activity into Proof Book

Part III - Practical Buyer Reality

- Chapter 8: "We Need a VDR-But Not Another Silo"
- Chapter 9: Standardize Secure External Collaboration (IT + Security)
- Chapter 10: Fundraising & Boards-Fast, Professional, Safe Enough

Appendices (tools you can use immediately)

- Appendix A: The VDR Outcome (one-page checklist)
- Appendix B: The 4-Phase Lifecycle (one-page operating model)
- Appendix C: Pilot Scorecard (how to prove outcomes in a pilot)

The promise of this book

By the end, you should be able to answer three questions with confidence:

1. **What VDR outcomes do we actually need (and how do we know we have them)?**
2. **How do we enforce those outcomes reliably inside Microsoft 365 (without creating another silo)?**
3. **How do we make it repeatable and defensible-so it scales beyond one deal?**

If this introduction resonates, you're the intended reader, and you're probably closer than you think. The hard part isn't finding another tool. The hard part is committing to an outcome standard and operating it consistently.

Part I

This part reframes how Virtual Data Rooms should be understood and evaluated. Instead of starting with tools, features, or vendors, it defines the outcomes organizations are actually trying to achieve when they say "we need a VDR."

You'll see why traditional VDRs work for narrow, one-off transactions but fail when collaboration needs to scale, repeat, or live inside the enterprise. The section concludes by laying out the core thesis of this book: how VDR-grade secure collaboration can be delivered natively inside Microsoft 365 without creating new silos, duplicating data, or breaking governance.

Best read if you are evaluating VDR options, questioning the cost and sprawl of existing tools, or trying to align business urgency with security and compliance realities.

Chapter 1: VDR Outcomes, Not VDR Tools

Most organizations don't want a "Virtual Data Room."

They want the **outcomes** a VDR is supposed to deliver - especially when sharing highly sensitive information with external parties.

That distinction matters because it changes how you evaluate solutions. If you treat a VDR as "a tool," you end up comparing feature lists. If you treat it as an "outcome standard," you focus on what the business truly needs: control, auditability, confidence, and a clean close-out.

This book starts with a simple reframing:

A VDR is not a product. A VDR is an outcome.

<u>1. What buyers mean when they say "We need a VDR."</u>

They are usually describing a consistent set of expectations. Those expectations exist **regardless of company size or deal type.**

Here's the translation.

Outcome 1: Controlled access to highly sensitive data

The expectation is that access can be tightly governed and restricted to specific people and groups.

Outcome 2: Secure collaboration with external parties

Buyers expect secure collaboration with external parties (investors, lawyers, auditors, partners) without falling back into email chains and uncontrolled file duplication.

Outcome 3: Clear visibility into who accessed what and when

Who accessed what and when - because the audit trail is part of the value.

Outcome 4: Confidentiality by default

It is not something added as an afterthought.

Outcome 5: Time-bound access and clean close-out

So, the organization can shut down or lock down access at the end of the process.

Outcome 6: External credibility

With an environment that feels professional and defensible for investors, boards, auditors, and legal stakeholders.

2. Why this outcome framing matters now

VDRs used to be treated as a 'one-off deal tool.' That's no longer the world most organizations operate in.

Your own buyer triggers show this clearly. Companies need these outcomes for:

- Active M&A/divestiture
- Fundraising/investor diligence
- Due diligence requests
- Audit or compliance review
- Board reporting with externals

In other words, sensitive external collaboration is recurring. The outcomes are needed repeatedly, across teams and departments - not just once.

That's why some buyers say:

"We need a VDR - but we don't want another silo, another security review, or unpredictable costs."

"Every deal or audit creates a new exception. We need a standard - not another one-off tool."

This shift defines the category:

From transactional VDR usage → to standardized secure external collaboration.

3. The thesis of this book (one sentence)

This book is built around a single category statement:

"Govern 365 delivers VDR outcomes inside Microsoft 365- without a standalone VDR data silo."

Even if you never buy any specific product, that sentence teaches a durable concept:

- The goal is VDR outcomes (not VDR tools).
- Those outcomes can be delivered *inside* Microsoft 365 (where many organizations already collaborate).
- And you avoid the friction, duplication, and exceptions created by a separate VDR repository ('data silo').

4. A simple mental model: Outcomes → Control → Repeatability

To keep the rest of the book practical, here is the simplest way to think about 'VDR-grade' collaboration:

A) Outcomes (what the business wants)

Controlled access, secure collaboration with external parties, visibility/auditability, confidentiality by default, time-bound access, clean close-out, and external credibility.

B) Control (how the outcomes are enforced)

Outcomes only matter if they are consistently enforced. This is where organizations need a reliable control layer: permissions discipline, protection discipline, and auditability discipline.

C) Repeatability (what makes it a 'standard')

The endgame is not a single perfect deal room. It's a repeatable pattern that can be reused across deals, audits, board work, and regulated collaboration - without creating a new 'exception' each time.

5. What this book will *not* do

This is not a product manual or a licensing guide.

The goal is to help readers:

1. Define the outcomes they need.
2. Understand why traditional approaches fail to scale into repeatable standards.
3. Adopt a lifecycle and operating model that makes VDR-grade collaboration defensible and repeatable.

That's why the next chapters move from 'outcomes' → to 'why traditional VDRs fail at scale' → to the Microsoft 365-native alternative and the lifecycle model that makes it defensible.

Chapter 1 Checklist: "Are we solving for outcomes or buying a tool?"

Use this checklist to pressure-test any VDR approach - vendor or home-grown. You should be able to answer these in under 15 minutes.

1. **Controlled Access:** Can we control who sees what (cleanly and consistently)?
2. **External Collaboration:** Can external parties collaborate without uncontrolled copies and email chaos?
3. **Auditability:** Can we answer "who accessed what and when" with confidence?
4. **Confidentiality Default:** Is confidentiality the default posture?
5. **Time-Bound Access:** Can we limit access to a time window and revoke cleanly?
6. **Clean Close-Out:** Can we close the room (lock down / archive / restrict) and keep it defensible?
7. **External Credibility:** Does it feel professional and credible to investors/boards/auditors?

8. **No New Silo:** Are we avoiding another repository and another security exception?
9. **Standardization Potential:** Can this become a reusable pattern across deals and departments (a 'standard')?
10. **Predictable Model:** Are we avoiding unpredictable costs and operational overhead?

If you can't answer these clearly, you're probably evaluating a VDR as a tool and not as an outcome.

Chapter 2: Why Traditional VDRs Fail at Scale

Traditional VDRs are not 'bad tools.' In fact, they often do exactly what they were designed to do.

They centralize documents, restrict access, and track downloads. They answer the question: **"Is the information present and secure?"**

The problem is that modern organizations no longer use data rooms as rare, one-off events. External collaboration is constant across deals, audits, fundraising, board reporting, and partner work.

When the need becomes continuous, traditional VDRs fail in a predictable way:

They solve one problem - and create others.

1) The Economic Model Doesn't Match How Work Scales

Traditional VDRs often carry a 'scarcity' pricing model: per-deal, per-transaction, per-user, per-page, or storage-based fees.

That model may look manageable when VDR usage is occasional. But once multiple stakeholders show up - legal, finance, advisors, auditors, board members - cost becomes hard to predict.

This is why the "VDR tax" narrative resonates: organizations feel like they are paying a premium to host and manage sensitive documents externally, and then paying again as usage expands or the deal lingers.

2) The Architecture Forces Data into a Separate Silo

Traditional VDRs typically mean a separate platform and duplicated content - data leaving corporate-approved repositories.

This matters because many organizations have already standardized security, identity, and compliance controls in Microsoft 365. The moment a deal room becomes an external portal, you create an exception: a separate perimeter, separate repository, and separate workflow.

This is exactly the question your security-oriented messaging asks:

"Why are we moving our most sensitive deal data out of Microsoft 365… in order to secure it?"

When the architecture requires data exfiltration into a third-party environment, the 'secure choice' starts to look like a risk tradeoff. Especially for organizations that already rely on Entra ID, Purview, and Microsoft audit/retention controls.

3) Tools built for one-off deals introduce friction that causes shadow IT

Traditional VDRs introduce friction that prevents repeatable collaboration

Traditional VDRs sit outside the day-to-day work patterns. Users must switch platforms, manage separate logins, and adapt to unfamiliar interfaces. That friction matters because deal teams need speed, IT needs standardization, and security needs control and auditability. Needs that are often already a source of concern. When a tool adds friction, teams route around it, reverting to email, shared links, and ad-hoc file sharing, which is exactly the behaviors organizations are trying to eliminate.

This problem is structural, not accidental. Traditional VDRs were designed for a world where deals were rare and temporary. In today's environment-where external collaboration is constant and higher-risk-they break down: they are deal-centric by design, difficult to govern across departments, and poorly suited to repeatable, scaled collaboration.

4) The "Close-Out" Problem: Deals End, But Risk Doesn't

A real VDR lifecycle includes closure: time-bound access, clean close-out, and defensible retention.

But traditional VDRs often make closure awkward:

- Data remains in the vendor portal after the deal is 'done.'
- Teams keep rooms alive for compliance reasons.
- Ongoing archive fees become a lingering cost and dependency.

In other words, the deal may close, but the platform dependency (and cost) can persist.

5) The Landscape in One Page: Three Approaches (and Why They Break)

Most organizations evaluating "a VDR" are choosing between three broad approaches:

A) Traditional VDR vendors

Strong security features and deal tooling, but commonly siloed data and expensive, complex pricing.

B) Modern/alternative VDRs

Often, better usability, but still standalone platforms requiring yet another repository and workflow.

C) Microsoft 365 alone (SharePoint/Teams)

The collaboration substrate is already there, but it's not always 'VDR-grade' out of the box without a repeatable governance pattern.

This book argues that the future is not 'a better standalone VDR.' It is a governed collaboration approach that lives where work already happens, and can be repeated as a standard.

6) The Practical Conclusion

Traditional VDRs remain useful for what they were originally built to do: a secure, time-bound portal around a static set of files.

But they fail when organizations try to scale the pattern across:

- multiple departments
- repeating diligence events

- modern security expectations
- continuous external collaboration

And that is why business teams increasingly expect 'this to live inside Microsoft 365.'

That shift sets up the next chapter.

If VDR is an outcome, and traditional VDRs don't scale, what does the modern alternative look like?

Chapter 2 Checklist: Signs You're Paying the "Traditional VDR Tax"

Use this checklist to spot the failure modes early.

1. You're paying for a separate platform to host files that already live in your corporate repositories.
2. Costs rise as soon as more stakeholders join (advisors, lawyers, auditors, board members).
3. The VDR requires manual uploads and the duplication of documents outside the normal workflow.
4. Teams route around the VDR and revert to email/links/ad-hoc sharing to move faster.
5. Every deal or audit feels like a new exception instead of a repeatable standard.
6. Closing a deal room leaves you dependent on vendor hosting and ongoing archive costs.
7. IT/security views the VDR as an architectural exception because sensitive data leaves the tenant boundary.

If multiple items are true, the issue is not that you chose the wrong VDR vendor. It's that the traditional VDR model is misaligned with how modern organizations operate.

Chapter 3: The Microsoft 365-Native Alternative (The Thesis)

This book's thesis is simple and intentionally repeatable:

"Govern 365 delivers VDR outcomes inside Microsoft 365-without a standalone VDR data silo."

That sentence matters because it describes a category shift - from renting a separate portal to **standardizing secure external collaboration inside the platform you already operate.** This model can be implemented in multiple ways; Govern 365 is one example of how organizations operationalize it.

1) What does "Microsoft 365-Native VDR" mean

A **Microsoft 365 Native Virtual Data Room** is a secure collaboration environment that lives entirely within your organization's own SharePoint Online inside your Microsoft 365 tenant. Unlike traditional VDRs, it leverages your existing investments in Microsoft security, identity, and compliance.

In practical terms, **the data room is not a separate place.** It's a governed collaboration pattern built on top of your Microsoft 365 environment.

2) The core idea: Microsoft 365 is the substrate. Govern 365 is the control layer.

Your internal lifecycle work captured the architecture framing clearly:

- Microsoft 365 serves as the **content and identity substrate.**
- Govern 365 acts as the **governance control plane** for provisioning, policy enforcement, evidence, and a seamless external experience.

That is the common alternative in one line: **keep the data where it already belongs and apply VDR outcomes in place.**

3) What stays the same (the outcomes)

The outcomes from Chapter 1 do not change. "VDR outcomes" still mean secure access, auditability, and confidence. The difference is **where and how** those outcomes are delivered:

- Same VDR outcomes
- No duplicated data store
- Data stays in your tenant
- No VDR silo

4) What changes in the architecture

A Microsoft 365-native approach eliminates the 'upload to the vendor' model.

Your sales and technical messaging state this explicitly:

- Zero Data Migration and 100% Data Sovereignty.
- Files stay in SharePoint (customer-owned), rather than a vendor cloud repository.
- The legacy risk is framed as: files leave your control, hosting fees apply, and duplicate data.

A native model is designed to keep the control signals (identity, protection, audit) in the same environment:

- Identity: Microsoft Entra ID (Azure AD) B2B guest access.
- Encryption & rights management: Microsoft Purview / Information Protection (including in-place encryption and restrictions like view-only/no download/no print).
- Audit trail: written to Microsoft Unified Audit Log (UAL) in the native architecture.

5) What changes (the security control model)

Traditional VDRs often rely on proprietary encryption models and separate security boundaries. A Microsoft 365-native model ties protection to the same identity system that grants access.

This is not "security theater." It is **policy enforcement attached to data and identity** in the environment that the organization already trusts.

6) What changes (the economics)

A key reason the Microsoft 365-native model is compelling is that it changes the cost structure:

- Flat subscription pricing instead of per-page/per-user/per-storage models.
- Unlimited users and storage, as positioned in the Microsoft-native vs traditional comparison table.
- Archiving is framed as "zero fees" with data staying in your tenant's account permanently.

7) What changes in terms of speed and adoption

A native approach reduces friction because it is integrated with the tools people already use:

- Native in Teams, SharePoint, and Office (rather than requiring manual uploads into a separate portal).
- Familiar M365 interface and "no training needed."
- Instant provisioning via automated templates.

That speed and familiarity matter because VDR usage is no longer limited to a single deal team. You're trying to scale secure external collaboration across corporate development, legal, finance/board, audits, and more.

8) A Microsoft 365-native VDR differs from traditional VDRs in the following ways:

- **Storage location:** vendor cloud vs your Microsoft 365 tenant.
- **Ownership:** vendor-controlled hosting vs you owning data, encryption keys, and policies.

Part II

This part introduces a simple, repeatable lifecycle for running sensitive external collaboration inside Microsoft 365. One that holds up under security review, audit scrutiny, and real operational pressure. Rather than treating a VDR as a temporary folder or a one-time deal room, it follows a four-phase model: setup, collaboration, evidence, and closeout. Each phase explains the decisions, controls, and guardrails required to make secure collaboration defensible during the project and after it ends.

Best read if you need to run live deals, audits, fundraises, or regulatory workflows today and want a pattern that works without heroics, exceptions, or constant IT intervention.

Chapter 4:
The 4-Phase Lifecycle That Makes It Defensible

Most virtual data rooms are treated as disposable. They are created for a single deal, used under pressure, and then abandoned exactly when governance, auditability, and retention matter most.

That disposal mindset is the root problem.

A VDR that is 'secure enough during the deal' but weak before and after it leaves organizations exposed to lingering access, missing evidence, and awkward reconstruction later. The lifecycle framework in this chapter exists to fix that.
It does so by applying a **repeatable, defensible operating pattern** across the full arc of sensitive external collaboration.

A defensible lifecycle is not about sophistication.

It is about predictability.

Specifically, a lifecycle that:

- starts with a controlled and intentional setup,
- sustains trust while collaboration expands,
- preserves evidence under negotiation and Q&A pressure,
- and closes with retention, proof, and controlled post-close access.

The Four Phases at a glance

The lifecycle is intentionally simple. Each phase answers one question:

1. Are we starting in control-or improvising?
2. Can collaboration expand without losing governance?
3. Can the room survive pressure without breaking?
4. Can we close cleanly and prove what happened?

If the answer is 'yes' at every stage, the lifecycle is defensible.

Phase 1: Pre-Fundraise (or Pre-Deal) & Preparation

Purpose: Decide whether everything that follows will be controlled or chaotic

Phase 1 determines the fate of the entire process. Most downstream problems, like guest sprawl, audit gaps, and messy close-out, can be traced back to how the room was created.

A defensible lifecycle starts with governed origination, not ad-hoc setup.

Instead of creating a folder, Phase 1 focuses on creating **action rooms from templates** that define:

- information architecture,
- default roles and ownership,
- external sharing posture,
- protection and retention intent.

Templates must do more than organize files so that the room is governed by design rather than discipline.

Defensibility outcome of Phase 1:

A room created through permissioned provisioning (not ad-hoc setup), with explicit owners and a consistent, template-driven security posture.

Phase 2: Investor Outreach & Engagement

(or External Sharing & Collaboration)

Purpose: Enable growth in participation without loss of control

Phase 2 is where most VDRs silently fail.

As external access expands to more investors, advisors, or reviewers, organizations often lose control without realizing it. Links multiply, versions drift, and visibility fades.

A defensible lifecycle treats external collaboration as **identity-bound and intentional**, not link-driven and reactive.

This phase emphasizes:

- identity-based access with time bounds and revocation,
- governed publishing and versioning (draft → approved → external),
- curated content bundles to prevent duplication and drift,
- and engagement telemetry that shows what is being accessed and by whom.

Just as importantly, the VDR must behave like a **collaboration environment**, not a static portal.

Defensibility outcome of Phase 2:

External collaboration that remains identity-governed and traceable, with intentional publishing patterns and measurable engagement signals.

Phase 3: Due Diligence & Negotiation

(The "pressure test" phase)
Purpose: Stay clean, attributable, and audit-ready under stress

This is where the lifecycle is tested the hardest.

Volume increases. Questions accelerate. Permissions change mid-stream. If governance weakens here, teams route around the room, and risk explodes.

A defensible Phase 3 supports:

- multiple parallel workstreams with ring-fenced access,
- unified control across item permissions, protection labels, and app-enforced restrictions (view-only, no download),
- and structured Q&A as a first-class, auditable workflow-not email threads.

The goal is not to control by friction. It is control that holds under pressure.

Defensibility outcome of Phase 3:

A negotiation environment where access, restrictions, Q&A, and evidence generation remain coherent without permission sprawl or off-platform workarounds.

Phase 4: Closing & Post-Fundraise
(or Close & Retain)
Purpose: Turn the end of the deal into the beginning of governance

Most VDRs treat "close" as the end.
A defensible lifecycle treats "close" as the moment governance becomes permanent.

This phase focuses on:

- clean access termination,
- intentional retention and disposition,
- and preserving proof for future inquiry.

Key needs typically include:

- modular e-signature integration with automatic labeling of signed artifacts,
- immutable closing binders with a verifiable chain of custody,
- governed post-close portals for reporting or ongoing access,
- and archives that support retention, legal hold, and sovereign export paths.

Defensibility outcome of Phase 4:

A close-out that produces controlled retention, governed post-close access, and defensible archives-not "turn it off and hope."

Cross-Cutting Capabilities

(The control plane that makes the lifecycle real)

The lifecycle only works if certain capabilities apply across all phases:

- **Policy as code:** templates encode naming, roles, labels, retention, guest lifecycle, and sharing posture.
- **Evidence & compliance:** immutable audit trails, exportable events, and attestation-ready reporting.
- **Tenant safety:** strict boundary enforcement, least-privilege access patterns, safe defaults.
- **Governed AI:** AI actions constrained to authorized content and fully auditable.
- **Semantic integrity:** metadata models and controlled vocabularies that prevent drift.

These are not extra features. They are what make repeatability possible.

The Practical Takeaway

This lifecycle is defensible because it aligns with how buyers actually evaluate risk, cost, and confidence-and because it applies beyond a single use case.

The same pattern applies across fundraising, M&A, audits, regulated collaboration, litigation, and boardrooms. That adaptability is the point: it turns VDR from a one-off deal tool into a repeatable enterprise capability.

Chapter 4 Checklist: *Is This Lifecycle Defensible?*

Use this checklist to pressure-test your approach:

1. **Governed creation:** Rooms are created through permissioned workflows with explicit owners and templates.
2. **Template discipline:** Templates define governance posture-not just folders.

3. **Identity-bound access:** External access is time-bound, revocable, and identity-based.
4. **Governed publishing:** Content moves through defined release states.
5. **Diligence-grade controls:** Permissions, protection, and Q&A remain coherent under pressure.
6. **First-class close-out:** Retention, disposition, and post-close access are designed, not improvised.

If you can answer "yes" to most of these, the lifecycle isn't just secure, but also defensible.

Chapter 5: Identity & External Access: Collaborate Without Losing Control

External collaboration is where VDR outcomes either become real or collapse.

Most organizations don't lose control because they chose the wrong folder structure. They lose control because **external access expands under pressure**: more investors, advisors, partner users, and "just send them the link" moments.

That is why this chapter is not about *how to invite guests*.

It is about how to design external access so that collaboration stays identity-bound, time-bound**, and revocable**-without slowing the deal down.

The Core Rule: External Access Must Be Identity-Bound (Not Link-Bound)

A defensible VDR lifecycle requires that external access defaults to **identity-bound access patterns**, with time bounds, continuous evaluation, and clean revocation. Low-friction alternatives can exist, but only as exceptions, not defaults.

This rule is the foundation of "collaborate without losing control":

- If a person's identity is not explicitly known, actions cannot be attributed.
- If access is not time-bound, it tends to linger.
- If revocation is fragile, close-out becomes fragile.

This is why the lifecycle model emphasizes identity-bound access and revocation as defaults-especially in Phase 2, when collaboration expands quickly and quietly.

Why One Access Model Is Not Enough

In theory, identity-bound access sounds simple. In practice, enterprises impose real constraints.

Some organizations support Microsoft Entra ID B2B guest access cleanly. Others restrict or block it-especially cross-tenant-due to regulatory, security, or architectural reasons.

A defensible external collaboration model must therefore support **two complementary modes of access**.

Access Mode A: Standard External Access (B2B Guest Access)

In this mode, external users are invited directly into the organization's Microsoft 365 tenant and granted access to the VDR workspace.

External users:

- receive a branded invitation,
- authenticate using their existing identity provider (Microsoft 365, Google Workspace, or supported OAuth providers),
- and access content through the same identity and permission model used internally.

What this gives you:

A clean, identity-bound path where access is granted, monitored, and revoked entirely within the tenant's existing governance boundary.

This is the preferred model wherever tenant policies allow it.

Access Mode B: Brokered External Access (for Restricted Tenants)

Some organizations cannot rely on direct guest access without derailing adoption.

For these environments, a brokered access path provides a practical fallback. One that preserves identity, accountability, and control even when cross-tenant restrictions exist.

What this gives you:

A way to support external collaboration without weakening tenant boundaries or forcing repetitive IT exceptions.

What 'Good External Onboarding' Looks Like

Good onboarding does two things at once:

- It minimizes friction for external users, and
- It preserves control for the owner.

A predictable onboarding flow typically looks like this:

1. External user receives an invitation with an access context.
2. Identity is established through an OTP-based verification and credential setup.
3. Access to the specific workspace is granted with tenant consent where required.
4. MFA is enforced.
5. The user sees only security-trimmed rooms and content they are authorized to access.

The specific mechanics matter less than the principle:

invitation → authentication → authorization → enforcement should be explicit, auditable, and repeatable.

The Real Governance Problem Is Not the 'Guests'-It's Guest Sprawl

External sharing failures usually follow a predictable pattern:

- links are shared too broadly,
- guest users remain long after the project ends,
- and teams lose visibility into who still has access.

The problem isn't that guests exist.
The problem is that the **guest lifecycle is unmanaged**.

This is why defensible external collaboration relies on three guardrails:

- controlled external sharing (who can share, and what they can share),
- automated access reviews so access doesn't linger indefinitely,
- and clear visibility through audit logs, reporting, and alerts.

Together, these reinforce the lifecycle model's insistence on time bounds and revocation as defaults, not cleanup activities.

What Must Be Standardized to Stay in Control

External access becomes repeatable only when a few things are standardized:

- Who can invite and under what conditions
- A clear invite → authenticate → authorize sequence
- A revocation posture that actually works when roles change

This is not a technical preference.
It is how you keep a VDR defensible when deal pace changes, stakeholders rotate, or pressure increases.

Without these standards, even strong tools degrade into improvisation.

Chapter 5:
Checklist: External Access That Stays Controlled

Use this checklist to assess whether your external collaboration model is defensible:

1. **Identity-bound by default**
   External access is identity-based (not link-based), with clear attribution.
2. **Time-bound and revocable**
   Access includes explicit time bounds and clean revocation as the default posture.

3. **Two access modes exist**
   You support both standard B2B guest access and a brokered option for restricted tenants.
4. **Predictable onboarding**
   The invite → authenticate → authorize → MFA flow is consistent and understandable.
5. **Guardrails are enforced**
   Controls exist over who can share externally and what can be shared.
6. **Guest sprawl is actively addressed**
   Access reviews and removal are built into the lifecycle-not treated as cleanup.
7. **Tenant constraints are respected**
   Allowed domains, guest settings, and external restrictions are handled intentionally.

If you can answer "yes" to most of these, you can collaborate externally without losing control, and the model will scale beyond a single deal.

Chapter 6:
Protection That Travels with the File

A VDR is not secure because it has a login screen.

It's secure because protection remains attached to the content even as it **moves**. When a file is downloaded, forwarded, copied, or opened outside the workspace.

That is what *"protection that travels with the file"* means in practice:

- the file is encrypted and rights-managed,
- only authorized people can open it,
- and the rules still apply after download or forwarding.

Without this, everything else looks controlled-until it isn't.

Why Access Control Alone Is Not Enough

Folder permissions and site access are necessary.

They are not sufficient for VDR-grade collaboration.

Once a document leaves the workspace-by download, forwarding, or copying-you need protection to ensure it persists *independently* of where the file lives. Otherwise, the most sensitive moment is the one you no longer control.

This is why defensible VDRs rely on **file-bound protection**, not just location-based access controls.

The Foundation: Sensitivity Labels Exist for This Exact Problem

Microsoft Purview describes sensitivity labels as a way to protect data *"as it travels inside and outside your organization."*

That definition matters.

Sensitivity labels are not just classification badges or compliance metadata. Used correctly, they are the mechanism that allows protection and enforcement to move with the content-across apps, devices, and sharing scenarios.

In VDR-grade collaboration, that capability is not optional.

The Core Control: Encryption That Travels with the File

The core protection pattern is simple and repeatable:

- **Trigger:** A VDR is created for a specific project or deal.
- **Automation:** A deal-specific sensitivity label is created.
- **Protection:** Files added to the VDR are automatically stamped and protected.
- **Result:** Even if a file is downloaded or forwarded, only authorized users can open it.

This is not 'security theater.' It is policy enforcement tied directly to identity and content-inside the environment the organization already trusts.

Why This Stays Defensible

Two design choices make this protection model reliable under pressure.

Location-Based Classification

Protection is tied to *where the file is placed*, not to user behavior.

Files inherit the correct label the moment they enter the governed workspace. Users are not responsible for manually classifying or protecting documents.

Granular, Deal-Scoped Encryption

Protection is scoped to the deal-not the entire company.

That allows each VDR to enforce its own access boundaries and usage rules, rather than relying on a blunt, enterprise-wide policy that is either too permissive or too restrictive.

Rights and Restrictions Are First-Class Controls

File-bound protection enables practical, visible controls that matter in high-stakes collaboration:

- encryption enforced at the item level,
- "View Only" access with no option to download or print,
- consistent behavior in the browser and supported clients.

Critically, forwarding does not create new access.
An unauthorized recipient may receive the file, but cannot open it.

PDFs matter and must be protected

Due diligence is full of PDFs.

A protection model that only works for Office documents is incomplete.

VDR-grade protection must apply to PDFs as well-otherwise, teams fall back to unprotected exports exactly when pressure increases.

If protection does not travel with PDFs, it does not travel far enough.

The Hard Problem Most Systems Ignore: Permissions Change

Real VDRs are dynamic.

People are added, removed, and moved between roles. Sometimes sharing actions alters effective permissions. Sometimes inheritance breaks.

Protection, therefore, cannot be a one-time action.

In a defensible model, file protection must remain aligned as access changes, so the security posture reflects current authorization rather than historical accident.

The Practical Payoff: Two Failure Modes Eliminated

When file-bound protection is implemented correctly, two common failure modes disappear:

1. **Possession does not equal access**
   Having a copy of a file is meaningless without authorization.
2. **Forwarding does not create uncontrolled distribution**
   Sharing an encrypted file does not grant access to new people.

This is the difference between *appearing secure* and *being defensible*.

Chapter 6 Checklist: Does Protection Really Travel with the File?

Use this checklist to pressure-test your protection model:

1. **Automatic protection**
   Files are automatically stamped and protected when added to the VDR.
2. **Download-resilient enforcement**
   Downloaded files remain encrypted and accessible only to authorized users.
3. **Granular rights**
   "View Only / No Download / No Print" restrictions are enforceable where required.
4. **Forwarding doesn't break security**
   Unauthorized recipients cannot open forwarded files.
5. **PDF protection is real**
   Protection applies to PDFs, not just metadata or headers.
6. **Permission changes stay aligned**
   Protection reflects current access, even as roles and permissions evolve.
7. **Purpose is explicit**
   The labeling strategy clearly exists to protect data *as it travels*.

If you can answer "yes" to most of these, you have VDR-grade protection.

Chapter 7: Evidence & Auditability: Turn Activity into Proof

In high-stakes collaboration, security is only half the job.

The other half is **proof**.

Proof matters because the hardest questions don't sound like security questions. They sound like legal, audit, and board questions:

- Who accessed the documents?
- What was shared-and when?
- Were permissions changed?
- Were files downloaded or deleted?
- Can we prove what existed in the room when the deal closed?

A VDR that cannot answer these questions in a defensible way forces organizations back into screenshots, email chains, and manual reconstruction. That isn't auditability. It's improvisation.

This chapter is about making **evidence-grade auditability** a core outcome of VDR-grade collaboration-not an afterthought.

Auditability Is Not a Report. It's an Evidence System.

Many tools treat auditability as a dashboard or a reporting feature.

A defensible VDR treats auditability as an **evidence system**:

- activity is captured,
- actions are attributable to identities,
- records are exportable,
- and evidence can be preserved as immutable proof at close-out.

Anything less may look informative, but it won't hold up under scrutiny.

The Practical Baseline: Auditability Inside the Tenant

Auditability must exist where the collaboration happens.

A defensible model provides two complementary layers of visibility:

- **At-a-glance, analytics** for each VDR, so owners and coordinators can see what's happening in real-time.
- **Deep, room-level analytics** for investigation, compliance, and post-event review.

Just as important is that this visibility must be available to the people who actually run the room, not only to administrators.

And it must be filterable along the dimensions that matter when questions arise:

- by user,
- by time,
- by activity type.

What "Evidence-Grade" Activity Looks Like

When activity is strong enough to become proof, it can be examined from multiple angles without reconstruction.

In practice, evidence-grade analytics typically include:

- KPI summaries (views, downloads, unique users, shares),
- trends over time,
- per-user engagement detail,
- internal vs. external activity splits,
- and full export capability.

This is what *"turning activity into proof"* means:

- activity is visible,
- it is sliceable,
- and it can be packaged directly for legal, audit, or board review.

Audit Reviews: A Discipline, not a Checkbox

Auditability becomes defensible only when it supports formal investigation-not just observation.

Within Microsoft 365, this means structured audit log review through the compliance portal, covering events such as:

- file access,
- sharing changes,
- permission updates,
- and deletions.

The distinction matters.

VDR-grade auditability is not simply about having logs.

It is **"we can execute a scoped investigation against defined event types."**

That capability is what separates a helpful report from defensible evidence.

The Step-Change: From Audit Trail to Evidence Archive

This matters most after the room is closed, when questions arrive without warning.

Logs alone are not always enough-especially after a deal closes.

A defensible close-out can produce an **evidence package** that captures:

- the content state,
- the permissions posture,
- and the activity record at a point in time,

with integrity that can be independently verified.

This transforms audit data from something you *query* into something you can *present*.

Why Evidence Reduces Both Risk and Friction

Strong evidence doesn't just reduce risk. It reduces friction.

External friction: "Trust me" becomes "Here's proof."

Instead of explaining verbally, you can show:

- what happened,
- when it happened,
- who performed each action,
- and what permissions were in effect.

This accelerates audits, reviews, and legal validation.

Internal friction: History doesn't need to be rebuilt

When evidence is built in, teams stop re-creating timelines after the fact. They spend less time reconstructing and more time deciding.

That alone can make or break a deal under pressure.

The Practical Takeaway

Evidence and auditability are not just 'nice-to-have reporting features.'

They are the mechanism by which VDR outcomes become **defensible**.

In a well-run VDR:

- activity is visible during the process,
- reviewable during diligence,
- and exportable into evidence at close-out.

That is how collaboration becomes proof.

Chapter 7 Checklist: Evidence-Grade Auditability

Use this checklist to pressure-test your audit model:

1. **Room-level analysis exists**
   Activity can be reviewed inside the VDR by user, date, and action.
2. **Owners can answer "who did what."**
   Views, downloads, uploads, and removals are attributable.

3. **Analytics are board- and audit-ready**
   Summaries, trends, and exports exist as first-class outputs.
4. **Formal investigations are supported**
   Audit searches cover access, sharing, permission changes, and deletion events.
5. **Close-out produces proof, not just an archive**
   Evidence packages include integrity verification and exportable artifacts.
6. **The evidence set is complete**
   Content snapshot, permissions state, activity log, and verification data are preserved together.

If you can check most of these boxes, you are no longer just tracking activity.
You are generating **defensible proof**.

Part III

This part focuses on scale. It shows how to move from successfully running a single VDR-grade workspace to standardizing secure collaboration across the organization. You'll learn how to communicate the same outcome-driven model differently to business leaders, IT, security, and compliance without changing the underlying architecture. The emphasis here is not on more control, but on shared language, repeatability, and trust, so sensitive collaboration becomes an enterprise capability rather than a special event.

Best read if your organization wants consistency across deals, departments, or regions and needs a way to institutionalize VDR-grade outcomes inside Microsoft 365 without slowing the business down.

Chapter 8: "We Need a VDR-But Not Another Silo"

Why this chapter exists (how it fits Part III)

Part II showed how to successfully run a single VDR-grade workspace. This chapter addresses what happens next-when that success needs to scale.

Once an organization proves it can handle one deal, audit, or fundraise inside Microsoft 365, the real question becomes broader and more strategic:

Can this approach become a standard-or does it create another special-case system?

This chapter helps readers recognize that the enterprise decision is not about features or tooling. It is about whether secure collaboration becomes an integrated capability or remains a series of exceptions.

You're Not Looking for a "Data Room Tool"

When organizations say that they need a VDR, they are rarely talking about software for its own sake.

They are trying to run a high-stakes process, like a transaction, diligence request, audit, board cycle, regulatory response, or fundraising round - where mistakes carry real consequences.

The requirement is simple to state and hard to execute:

- Share highly sensitive information with external parties
- Move quickly without improvisation
- Maintain confidence that the activity can be explained and defended later

This is not a theoretical problem. The timeline, pressure, and the outcome matter.

What's Already True Inside the Organization

For most organizations, the center of gravity has already shifted.

Teams live inside Microsoft 365:

- Work happens in Teams
- Files live in SharePoint and OneDrive
- Documents are created, reviewed, and revised in Office apps

What has changed is not collaboration itself.
What has changed is the **volume, frequency, and risk of external collaboration**.

Email, ad-hoc sharing, and 'temporary' links no longer hold up.
Especially when:

- External parties change
- Timelines extend
- Regulators or auditors ask questions months later

This is usually the moment when one of these triggers appears:

- "We need a VDR for an active deal or project."
- "Our current VDR is too expensive or hard to manage."
- "We want secure external collaboration inside Microsoft 365."
- "We failed an audit or security review."
- "We need more visibility and control."

At this point, the organization is already in motion.

The choice is no longer abstract.

What "We Need a VDR" Actually Means

When buyers say they need a VDR, they are not asking for a feature set.

They are asking for a specific set of outcomes:

- Controlled access to highly sensitive data
- Secure collaboration with external parties
- Clear visibility into who accessed what and when
- Confidentiality by default
- Time-bound access with a clean close-out
- External credibility-the process must feel professional and defensible

That is the real requirement.

Everything else is secondary.

The Constraints That Shape the Decision

At scale, the decision is defined more by constraints than preferences.

Organizations consistently draw the same lines:

1. Avoid creating a new data silo

Moving the most sensitive information into a separate vendor system introduces duplication, ownership questions, and governance gaps. Teams increasingly expect sensitive work to remain within the enterprise controls that already exist.

2. Avoid resetting security for every process

Re-establishing identity, permissions, and controls from scratch for each deal slows execution and increases risk. Organizations want the existing Microsoft 365 security posture to remain the foundation.

3. Avoid unpredictable cost curves

Per-page, per-user, or per-deal pricing models break down as stakeholder lists grow and timelines stretch. When external counsel, auditors, or additional bidders are added, cost should not become a surprise.

These constraints don't eliminate options; they clarify what is acceptable.

The Real Decision Being Made

Stripped of vendor language, the enterprise decision is straightforward:

Do we keep sensitive collaboration within Microsoft 365 and apply VDR-grade outcomes, or do we export our most sensitive data to a standalone system to achieve those outcomes?

This question explains why comparisons around:

- Data location
- Ownership
- Pricing model
- Access controls
- Auditability
- Close-out behavior

All these matters far outweigh feature checklists.

They map directly to the risks and responsibilities the organization already carries.

What 'Success' Actually Looks Like

Successful implementation is not defined by spinning up a room.

It looks like this:

1. Teams can share sensitive material externally without improvising.
2. External participants can collaborate without confusion or duplication.
3. Activity can be answered confidently: who accessed what, and when.
4. When the process ends, access closes cleanly-and evidence is retained appropriately.

This is why Appendix A exists: it provides a concise outcome checklist to pressure-test whether a solution delivers a defensible standard or merely adds another tool.

How to Evaluate Any Solution (Reader-First Questions)

Regardless of vendor, the same questions apply:

1. Where does the data live during the process?
2. Who owns the controls, policies, encryption, and access?
3. Can it deliver full VDR outcomes, not just secure file sharing?
4. What happens when the stakeholder list grows?
5. Can activity be reviewed and exported later as evidence?
6. Is it a close-out of a designed phase, or an improvised cleanup?
7. Is the external access role-based, auditable, and revocable?
8. Can this pattern be repeated without starting over each time?

If these answers aren't clear, the solution is solving the wrong problem.

The Simplest Way to Start

Scaling does not require over-design.

A practical pilot proves one thing only:

Can a real process be run end-to-end in a controlled, repeatable way?

A clean pilot includes:

- A clearly defined workspace and owner
- A repeatable template
- Explicit roles and protections
- External participation
- Active monitoring
- Clean close-out and retention

That is enough to determine whether VDR-grade outcomes can exist without creating another silo.

Chapter 9: Standardize Secure External Collaboration (IT + Security)

Standardizing Secure External Collaboration (IT + Security)

Chapter 8 focused on completing a single high-stakes process without creating another silo.

Chapter 9 addresses the next, larger problem: preventing the organization from living in exception-land.

At this stage, the issue is no longer whether secure external collaboration *can* work inside Microsoft 365. It's whether it can become the standard way the organization operates-instead of a special case solved repeatedly with new tools, new rules, and new risks.

This chapter is written primarily for IT, Security, and Compliance leaders-but it is equally relevant to anyone responsible for enterprise consistency.

External Collaboration Is No Longer an Edge Case

External sharing is no longer occasional. It is continuous.

Deals, audits, partners, regulators, and board interactions now overlap-and each one introduces the same underlying risks. As volume increases, the failure pattern becomes predictable:

- External sharing policies drift from site to site
- Visibility degrades ("who still has access?" becomes hard to answer)
- Sensitive data leaves Microsoft 365 without consistent protection
- Permissions sprawl across dozens-or hundreds-of workspaces

This is why 'standard Microsoft 365 sharing' often feels unsafe at scale.

The issue is not the platform but the inconsistent application, monitoring, and cleanup.

The Outcome IT and Security Are Trying to Achieve

The goal is not a better deal room.

The goal is **standardized, policy-driven external collaboration inside Microsoft 365**.

That means:

- External sharing is governed consistently
- Access does not linger or sprawl
- Auditability exists without manual reconstruction
- Sensitive data stays within the organization's control boundary

Without standardization, business teams need speed, IT needs consistency, and security needs visibility. But all three work against each other. Standardization is what allows them to align.

When the IT-Led Standardization Decision Appears

IT, Digital Workplace, or Compliance leaders typically enter the picture when:

- Tool sprawl increases (multiple VDRs alongside Box, Dropbox, ShareFile, etc.).
- A security incident or audit exposes gaps.
- Deal and audit volume increases beyond 'occasional.'
- Leadership mandates external sharing standardization.

The internal conclusion is usually stated plainly:

"Every deal or audit creates a new exception. We need a standard, not another one-off tool."

The Actual Decision Being Made (IT Perspective)

From an IT standpoint, the choice is not about features.

It is a decision between:

1. Continuing to allow sensitive external collaboration through a mix of ad-hoc sharing and disconnected tools, or,

2. Establishing a repeatable external-collaboration standard inside Microsoft 365 that business teams can use without creating governance exceptions.

The second option requires more than a place to upload files. It requires policy-based controls, automated lifecycle management, and continuous visibility.

What 'Standardization' Really Means in Practice

Standardization is successful when these questions can be answered without heroics:

- Where is sensitive external collaboration allowed, and where is it not?
- Who has access right now, and why?
- What happens when projects end? Do guests linger or close out cleanly?
- Can activity be reviewed and explained later without forensic reconstruction?

This is why standardization must combine **controls, automation, and monitoring**, not manual enforcement and best intentions.

How IT Evaluates Solutions

Evaluation criteria are consistent across organizations:

- Does this reduce tool sprawl or add to it?
- Can it be applied consistently across departments?
- Can business teams operate without opening a ticket for every exception?
- Can oversight be operationalized without custom reporting?

If standardization depends on constant intervention, it will not scale.

What Success Looks Like (IT-Defined)

From an IT standpoint, success is not deployment.

Success looks like:

- Fewer exceptions
- Fewer surprises
- A repeatable pattern business teams understand
- Audit readiness without scrambling

At that point, secure external collaboration is no longer a risk multiplier. It becomes an operational capability.

The Security-Led Perspective

Security teams often enter the conversation with a different question:

"Why are we moving sensitive data out of Microsoft 365 in order to secure it?"

Security-led buyers typically engage when:

- Data sovereignty is questioned
- Third-party clouds are rejected
- Audit findings escalate concerns
- The CISO becomes directly involved

These buyers are not persuaded by features.
They want proof that risk has been reduced without disrupting **business execution**.

What Security Needs to See to Say "Yes"

From a security perspective, success means:

- Lower risk of oversharing
- Reduced lingering guest access
- Clear visibility into access and activity
- A defensible audit trail that does not rely on a vendor's black box

Security is not trying to stop collaboration.
It is trying to ensure that collaboration behaves predictably under scrutiny.

One Chapter, Two Buyers, One Outcome

IT and Security arrive from different angles, but reach the same conclusion:

Secure external collaboration must become a **repeatable enterprise standard inside Microsoft 365 - not another exception**.

This is the shift from:

- Transactional VDR usage to
- Standardized secure external collaboration

The underlying operating model remains the same. What changes is scale, consistency, and accountability.

Practical Self-Check: Are You in Chapter 9 Territory?

Use these questions to assess whether your organization has entered the standardization phase:

1. Are external-sharing policies inconsistent across sites today?
2. Is visibility into external access limited?
3. Does sensitive data leave without consistent protection?
4. Does the same problem repeat across deals, audits, and partners?
5. Is the goal to standardize inside Microsoft 365 rather than add another tool?

If most answers are "yes," the organization is no longer buying a VDR.

It is establishing a standard operating model for secure external collaboration.

Chapter 10: Fundraising & Boards-Fast, Professional, Safe Enough

Fundraising & Boards-Fast, Professional, Safe Enough

This chapter is about urgency, not transformation.

You are not trying to modernize governance.

You are not trying to standardize collaboration across the enterprise.

You are trying to get something important done-**quickly**-without embarrassing yourself or creating unnecessary risk.

This chapter applies when:

- you are raising capital,
- preparing board materials,
- responding to investor diligence,
- or sharing sensitive financial or strategic information with people who don't work for you,

And the clock is already ticking.

What's Actually Happening in these Moments

Under pressure, most teams default to whatever feels fastest:

- email attachments,
- shared links,
- ad-hoc folders,
- or a rushed signup for a third-party VDR.

This works until it doesn't.

These are also the moments when mistakes carry disproportionate consequences:

- an investor forwards a deck,
- a board document leak,

- someone retains access longer than intended,
- or months later, no one can answer "who saw what?"

You don't need enterprise theory here.

You need something that immediately feels legitimate, controlled, and professional.

The Real Job-to-Be-Done

When founders and executives say "we need a data room at this moment," they are not asking for features.

They are asking for reassurance:

- that sensitive material isn't floating around uncontrolled,
- that external parties have professional experience,
- and that nothing obviously reckless is happening behind the scenes.

The real job-to-be-done is simple:

Make this sharing process feel legitimate to investors, to the board, and to ourselves.

What 'Safe Enough' Means Here

This chapter is intentionally not about maximum security or long-term operating models.

It is about preventing obvious, avoidable mistakes without slowing the process down.

'Safe enough' usually means:

- access can be controlled,
- access can be removed,
- documents aren't casually forwarded or downloaded,
- and there is basic visibility into engagement.

It does **not** mean:

- building an enterprise governance framework,

- designing cross-department standards,
- or involving multiple layers of IT approval.

Those things may come later if and when the organization needs them.

How Decisions Are Made in this Moment

Decisions here are pragmatic and time-compressed.

You care about:

- speed to set up,
- how it looks to external stakeholders,
- and whether it avoids obvious risk.

You want to avoid:

- long procurement cycles,
- complex configuration,
- or anything that forces your team to relearn how to work.

This is why familiar Microsoft 365 tools matter so much in this scenario, as it is the shortcut to confidence.

What You're Actually Deciding

The real decision is not:

"Which VDR vendor has the most features?"

It is:

"How do we share sensitive materials quickly, without losing control or credibility?"

That decision usually comes down to three implicit tests:

1. Does it look professional to investors or board members?
2. Does it reduce obvious sharing risk?
3. Can access be shut down cleanly when this is over?

If the answer to those is "yes," the solution is good enough for this phase.

What Success Looks Like from Your Point of View

Success here is uncomplicated:

- investors can access what they need,
- board members receive materials securely,
- nothing leaks,
- and when the moment passes, access can be closed cleanly.

No one praises the system.

No one complains.

You move on to the next problem.

That's the win.

When This Turns into Something Bigger

Many organizations first encounter controlled data rooms during fundraising or board work.

Later, they realize the same pattern applies to:

- M&A,
- audits,
- legal matters,
- regulatory collaboration.

But this chapter is not about that future.

It is about solving today's problem without over-engineering tomorrow's.

If the organization requires additional structure, the fundamental outcomes-such as controlled access, auditability, and a clean closeout-remain scalable.

For now, the goal is simpler:

Share sensitive information responsibly, without slowing down the business.

Appendix A – The VDR Outcome

Most organizations don't actually want a "Virtual Data Room." They want the outcomes a VDR is supposed to deliver-especially when sharing highly sensitive information with external parties.

Use this as a single-page "VDR-grade" requirements checklist.

A) Outcome Standard (what the business expects)

Check "Yes" only if your approach consistently delivers the outcome.

1. **Controlled Access**
   Can we control who sees what-cleanly and consistently-when sharing highly sensitive data?
2. **Secure Collaboration with External Parties.**
   Can external parties collaborate without uncontrolled copies, email chaos, or ad-hoc sharing workarounds?
3. **Clear Visibility / Auditability**
   Can we see (and later prove) who accessed what and when?
4. **Confidentiality by Default**
   Is confidentiality the default posture-rather than something users must remember to enforce manually?
5. **Time-Bound Access + Clean Close-Out**
   Can we enforce time-bound access and achieve a clean close-out (end the project without leaving loose access behind)?
6. **External Credibility**
   Does the experience create external confidence (investors, bidders, auditors, counsel) that the process is controlled and professional?

B) Control Layer (how outcomes are enforced, not promised)

Outcomes only matter if they are consistently enforced:

1. **Identity-Anchored Access (not link-only sharing)**
   Does access tie back to identity controls (so actions are attributable), rather than anonymous distribution patterns?
2. **Protection That Travels with the File**
   If a file is downloaded, does protection still apply so that only authorized users can open it?
3. **Granular Rights (view-only / restricted use when needed)**
   Can you enforce granular rights, such as view-only/no download/no print, directly in the browser when required?
4. **Audit Trail Is Native and Durable**
   Is the audit trail written into a durable system of record (not a vendor-only log you can't operate)?
5. **Actionable Reporting Exists (not just "we have logs")**
   Are reports available that cover workspace/VDR analytics and activity audit (not just raw logs)?

C) Repeatability (what makes it a "standard," not a one-off room)

The endgame is not one perfect deal room. It's a repeatable pattern you can reuse across deals, audits, board work, and regulated collaboration without creating a new exception each time.

1. **Repeatable Setup (templates + governed provisioning)**
   Can you create VDRs through a controlled provisioning model (templates, required owners, and consistent configuration)?
2. **No Silo Requirement (stays aligned to Microsoft 365)**
   Do you deliver VDR outcomes-secure access, auditability, and confidence-without a standalone VDR platform?
3. **"Native user experience" principle**
   Does the approach preserve a native sharing/permissions experience so users can use the software they know and love, rather than forcing them to manage new abstractions?

How to use this appendix (quick scoring)

- If you score “Yes” on **A1–A6**, you have the *outcome definition* nailed.
- If you score “Yes” on **B1–B5**, your outcomes are *enforceable*.
- If you score “Yes” on **C1–C3**, you’ve moved from a “deal tool” to a *repeatable standard*.

Appendix B – The 4-Phase Lifecycle

A defensible lifecycle is not 'more features.' It is a predictable operating pattern that starts with controlled setup, sustains trust during collaboration, preserves evidence during negotiation and Q&A, and closes with retention, proof, and controlled post-close access.

This lifecycle framework was reviewed and validated as solid and aligned to how buyers budget and evaluate VDR outcomes. It is designed to make VDR-grade collaboration **repeatable and defensible** across fundraising, M&A, audits, regulated collaboration, and boardrooms.

Phase 1 - Pre-Fundraise (or Pre-Deal) & Preparation

Purpose: Decide whether everything that follows is controlled or chaotic.

What must be true in Phase 1?

- Create *action rooms from templates* that define information architecture, labels, retention, external sharing posture, and default roles.
- Templates must include more than folder structure: a metadata model, publishing states, and an approval workflow so the room is governed, not just organized.
- Where required, **NDA and accreditation gates** should be implemented as identity/access gating workflows with **immutable evidence** written to the audit trail.

Phase 1 checklist:

- Room created from a template with defined governance posture (roles + sharing posture + retention).
- Owners and default roles are set and consistent.
- Publishing workflow exists (draft → approved → external-visible) rather than ad-hoc sharing.

Phase 2 - Investor Outreach & Engagement

Purpose: Enable safe distribution and engagement without losing control.

What must be true in Phase 2?

- Access patterns should default to identity-bound access with time bounds, continuous evaluation, and revocation, with clear alternatives for low-friction access when needed.
- Version control should operate as a **governed release channel** (internal drafts → approved publications → investor-visible major versions), with notifications that carry evidence.
- *Bundles* (curated sets) should be driven by metadata/publishing rules to prevent duplication and drift.
- Engagement telemetry should be comprehensive, including document views, dwell time, exports, anomalies, and executive summary reporting.

Phase 2 checklist:

- External access is identity-bound and revocable, not 'link sprawl.'
- External-visible content is governed by publishing states and version discipline.
- Engagement tracking exists at the level investors/buyers care about (views/dwell/exports).

Phase 3 - Due Diligence & Negotiation

Purpose: Keep permissions + protection + evidence coherent under pressure.

What must be true in Phase 3?

- Support multiple parallel workstreams with **ring-fenced permissions**, and optionally 'clean room' modes.
- Granular permissions must remain unified across: SharePoint item permissions, sensitivity labels, and app-enforced restrictions such as view-only and disabled download.

- Q&A should be a first-class module: structured, role-based, and auditable (triage, assignment, review/approval).
- Dynamic watermarking becomes a meaningful deterrent when applied consistently across view/export/print with identity + timestamp evidence.

Phase 3 checklist:

- Permissions remain coherent as parties change (no uncontrolled exception paths).
- Q&A is captured as an auditable workflow, not email threads.
- Protection model holds even under pressure (e.g., "encryption travels with the file").

Phase 4 - Closing & Post-Fundraise (or Post-Deal)

Purpose: Close-out is first-class: retention, proof, and controlled post-close access are designed-not improvised.

What must be true in Phase 4?

- E-signature integration should be modular (common providers) with automatic labeling/retention of signed artifacts.
- Closing binders should be immutable, provenance-rich, and exportable with a chain of custody.
- Post-close portals become governed workspaces for reporting packs/announcements with controlled ongoing access.
- Archives should include retention label application, legal hold support, disposition review, and customer sovereign export paths.
- Lifecycle governance signals matter: guest recertification and disposition are explicitly called out as part of 'govern the data beyond the deal.'

Phase 4 checklist:

- Evidence/closing artifacts are retained with provenance and exportability.
- Access is reduced/controlled post-close (not abandoned).
- Retention/disposition is deliberate (labels/holds/review) rather than 'delete the room and hope.'

Cross-Cutting Capabilities (apply across all phases)

These are the make-it-explicit capabilities called out in the lifecycle review:

- **Policy as code for templates:** naming, roles, labels, retention, DLP, guest lifecycle, access reviews, external sharing posture.
- **Evidence & compliance:** immutable audit trails, event export APIs, attestation reports aligned to common frameworks.
- **Tenant safety:** strict tenant boundary enforcement, least-privilege Graph access, per-tenant keys, safe defaults for external collaboration.
- **Multi-plane AI (governed):** central governance model + sovereign execution paths, constrained to authorized content and auditable AI actions.
- **Semantic integrity:** metadata model, controlled vocabularies, classification, prevention of drift across bundles/workstreams.
- **Calm enterprise UX:** minimal clicks, clear status states, guided flows, safe reversible actions for access changes and publishing.

"If you only remember one thing."

Most VDRs are treated as disposable; they are created, used under pressure, then abandoned exactly when governance, auditability, and retention matter most. This lifecycle exists to prevent that by making VDR-grade collaboration governed across the full arc of sensitive external collaboration.

Appendix C – Pilot Scorecard

This scorecard is designed to answer one question clearly:

Did the pilot prove 'VDR outcomes' and not just features? And can this become a repeatable standard?

A VDR pilot is successful only if it demonstrates the outcome standard buyers expect: controlled access, secure collaboration with external parties, clear visibility into who accessed what and when, confidentiality by default, time-bound access + clean close-out, and external credibility.

0) Pilot Charter (define what 'success' means before you start)

Pilot type (pick one):

- **Deal / Fundraising pilot** (business-initiated)
- **Standardization pilot** (IT-initiated)
- **Security proof pilot** (CISO-initiated) *(Pilot types align with the "VDR outcomes / no silo" framing and stakeholder model in the pitch.)*

Pilot scope (must be explicit):

- One real room (deal, diligence, board packet, or fundraising) with a defined external party list.
- The room must be created using the governed setup motion (template/owners/role model).

Owners (must be named):

- Primary Owner + Secondary Owner (avoid orphaned rooms).

Success definition: A "GO" is when the pilot demonstrates VDR outcomes *and* shows the lifecycle is defensible end-to-end (not just secure during the deal).

1) Entry Gate (Pilot "READY" checklist)

A) Provisioning & template readiness (repeatability starts here)

- Room creation permission is controlled by a dedicated permission group. Not everyone can create rooms.
- Room creation follows: Deal Room Type → Configuration Form → Template-based provisioning.
- Template includes pre-configured permissions/features (roles, initial structure, default security settings, and linked lists such as Q&A tracker, where applicable).

B) Deployment/validation expectation is set (avoid false failure)

- Stakeholders understand that full protection readiness can require a validation window for Microsoft 365 propagation (labels and related configurations may take 1–2 days to fully propagate).

The pilot cannot start until these are true - otherwise, the pilot measures 'setup friction' rather than outcomes.

2) Scorecard (Outcome Proof)

Use this scoring method:

- **PASS** = demonstrably true in pilot (with evidence artifact)
- **CONDITIONAL** = works with limitations; limitation is documented + accepted
- **FAIL** = not demonstrated / breaks under real usage

A) Outcome Proof (what the buyer expects)

1. **Controlled Access** - can you control who sees what?
2. **Secure External Collaboration** - can externals collaborate without uncontrolled copies/email chaos?
3. **Auditability / Visibility** - can you show who accessed what and when?
4. **Confidentiality by Default** - confidentiality posture is the default expectation.

5. **Time-Bound Access + Clean Close-Out** - can you end access cleanly and intentionally?
6. **External Credibility** - does the process feel defensible and 'real' to externals?

B) Control Proof (the enforcement layer must hold under pressure)

1. **Protection travels with the file** - downloaded files remain protected; only authorized users can open them (the "encryption travels with it" story).
2. **Lifecycle governance signals** - the product supports governance beyond a one-off room (recertification /disposition posture is part of the lifecycle story).
3. **Evidence-grade close-out (if in pilot scope)** - ability to close with proof (archive workflow/evidence package) where applicable.

3) Lifecycle Coverage Score (Phase-based: did the pilot prove the operating pattern?)

The lifecycle is meant to be defensible across the full arc, not just during the deal.

Score each phase as **PASS / CONDITIONAL / FAIL / NOT TESTED** based on whether you produced the listed artifacts.

Phase 1 - Pre-Deal & Preparation (governed setup)

Evidence artifacts example:

- Template-based 'action room' configuration (architecture /labels/retention/sharing posture/default roles).

Phase 2 - Outreach & Engagement (safe distribution + telemetry)

Evidence artifacts example:

- Identity-bound access pattern + time bounds/revocation concept documented.
- Engagement telemetry (views, dwell time, exports/anomalies, executive summary reporting) demonstrated or captured.

Phase 3 - Diligence & Negotiation (pressure test)

Evidence artifacts example:

- Ring-fenced workstreams; unified permissions across item permissions + labels + app restrictions.
- Structured, role-based Q&A evidence if used.

Phase 4 - Close & Post-Close (retain, prove, control)

Evidence artifacts example:

- Defined close-out posture, retention/hold/disposition approach, and controlled post-close access.

4) 30/60/90 Review Cadence (recommended)

Note: The sources define the lifecycle and outcomes, but they do not prescribe a specific 30/60/90 cadence. The cadence below is a recommended operating rhythm to keep the pilot measurable and on track.

Day 30 - "Outcome Proof"

Goal: prove the Outcome Standard (A1–A6) in one real room:

- Outcome Proof section completed with evidence artifacts (screenshots/exports).

Day 60 - "Control Under Pressure"

Goal: prove enforcement (protection that travels, auditability under real usage).

- Show at least one 'pressure' scenario: role changes, folder access change, external user onboarding/removal, and audit traceability.

Day 90 - "Repeatability Decision"

Goal: decide whether this becomes a standard (templates + lifecycle pattern).

- Phase scorecard completed for the phases tested; gaps explicitly listed as next build/config items.

5) Final Gate Recommendation (what you report to leadership)

Choose one:

- **GO** - Outcomes proven + lifecycle mostly covered + repeatability path is clear. Re: VDR Lifecycle discussion
- **CONDITIONAL GO** - Outcomes proven, but specific lifecycle/control gaps remain; proceed with constraints and a remediation plan.
- **NO-GO** - Outcomes not proven or controls break under real usage; do not scale until blockers are resolved.

Required attachments to the recommendation:

- Outcome Proof checklist (A1–A6) with evidence artifacts.
- Lifecycle phase score (Phase 1–4) with "Not Tested" explicitly marked where applicable.
- Reporting evidence (filterable activity report export) if auditability is claimed.

Acknowledgements

This book could not have been completed without the help of many of my colleagues who helped shape my thinking. Here's a few to thank personally:

- Viraj Bais, Co-founder and CTO at Netwoven and a long time friend
- Mark Ferraz, Senior Technical Director and Product visionary of Govern 365
- Navoneel Talukdar, Technical Architect and core product delivery team member of Govern 365
- Debashis Mandal, Technical Architect and core product delivery team member of Govern 365
- Sujay Ghatak, Customer Success Manager of Govern 365
- Andrew Sutton, VP of Solution Sales at Netwoven for being the early visionary of Govern 365
- Matt Maher, VP of Dellivery at Netwoven
- Subhendu Das, Sr. Technical Director at Netwoven
- Gursimran Singh, Principal Consultant at Netwoven for the early sales of Govern 36

There are many others that have contributed to this thinking at Netwoven, Microsoft, and our customers. Thank you all.

About the Author

Niraj Tenany is the Chief Executive Officer of **Netwoven**, a Microsoft-focused technology consulting and solutions firm specializing in secure collaboration, data governance, and enterprise digital transformation. He has spent his career working at the intersection of business outcomes, security, and operating discipline-helping organizations modernize without creating unnecessary risk or complexity.

Niraj works closely with business leaders, IT teams, and security stakeholders to operationalize collaboration inside Microsoft 365 in ways that are scalable, defensible, and repeatable. His work spans sensitive external collaboration, identity and access governance, compliance-driven workflows, and AI-ready information architectures for regulated industries.

As the product owner of **Govern 365**, Niraj has been directly involved in designing and delivering Microsoft 365-native Virtual Data Room outcomes-secure access, auditability, and confidence without relying on standalone data room platforms or external silos. The frameworks and lifecycle models in this book are informed by real-world implementation, customer pressure, and post-event accountability.

Niraj lives in California and writes for leaders who are responsible not just for getting work done, but for being able to explain and defend how it was done later.

www.ingramcontent.com/pod-product-compliance
Lightning Source LLC
LaVergne TN
LVHW050923210926
848482LV00012B/511
9798901904633